Hazardous Materials Accident Report

Release and Ignition of Hydrogen Following Collision of a Tractor-Semitrailer with Horizontally Mounted Cylinders and a Pickup Truck near Ramona, Oklahoma May 1, 2001

National Transportation Safety Board Washington, D.C.

this page intentionally left blank

Hazardous Materials
Accident Report

Release and Ignition of Hydrogen Following Collision of a Tractor-Semitrailer with Horizontally Mounted Cylinders and a Pickup Truck near Ramona, Oklahoma May 1, 2001

NTSB/HZM-02/02
PB2002-917003
Notation 7371A
Adopted September 17, 2002

National Transportation Safety Board
490 L'Enfant Plaza, S.W.
Washington, D.C. 20594

National Transportation Safety Board. 2002. *Release and Ignition of Hydrogen Following Collision of a Tractor-Semitrailer with Horizontally Mounted Cylinders and a Pickup Truck near Ramona, Oklahoma, May 1, 2001.* **Hazardous Materials Accident Report NTSB/HZM-02/02. Washington, DC.**

Abstract: On May 1, 2001, a tractor-semitrailer that had horizontally mounted cylinders filled with compressed hydrogen struck a pickup truck near Ramona, Oklahoma. According to witnesses, the tractor-semitrailer then went out of control and overturned while continuing along the highway. It went off the road to the east and traveled 300 more feet before it stopped. During the process, some of its cylinders, valves, piping, and fittings were damaged and released hydrogen..

The safety issues discussed in this report are the adequacy of Federal requirements for protecting cylinders that are horizontally mounted on semitrailers and the valves, piping, and fittings of the cylinders during rollover accidents and the adequacy of the information in the *North American Emergency Response Guidebook* about compressed hydrogen

As a result of its investigation of this accident, the Safety Board made recommendations to the Research and Special Programs Administration.

The National Transportation Safety Board is an independent Federal agency dedicated to promoting aviation, railroad, highway, marine, pipeline, and hazardous materials safety. Established in 1967, the agency is mandated by Congress through the Independent Safety Board Act of 1974 to investigate transportation accidents, determine the probable causes of the accidents, issue safety recommendations, study transportation safety issues, and evaluate the safety effectiveness of government agencies involved in transportation. The Safety Board makes public its actions and decisions through accident reports, safety studies, special investigation reports, safety recommendations, and statistical reviews.

Recent publications are available in their entirety on the Web at <http://www.ntsb.gov>. Other information about available publications also may be obtained from the Web site or by contacting:

National Transportation Safety Board
Public Inquiries Section, RE-51
490 L'Enfant Plaza, S.W.
Washington, D.C. 20594
(800) 877-6799 or (202) 314-6551

Safety Board publications may be purchased, by individual copy or by subscription, from the National Technical Information Service. To purchase this publication, order report number **PB2002-917003** from:

National Technical Information Service
5285 Port Royal Road
Springfield, Virginia 22161
(800) 553-6847 or (703) 605-6000

Contents

Executive Summary

About 2:15 p.m., central daylight time, on May 1, 2001, a northbound tractor, in combination with a semitrailer that had horizontally mounted cylinders filled with compressed hydrogen, which is a flammable gas, struck a northbound pickup truck that had veered in front of the tractor-semitrailer on U.S. Highway 75, 2 miles south of Ramona, Oklahoma. According to witnesses, the tractor-semitrailer then went out of control and overturned while continuing along the highway. It went off the road to the east and traveled 300 more feet before it stopped. During the process, some of the cylinders, valves, piping, and fittings at the rear of the semitrailer were damaged and released hydrogen. The hydrogen ignited and burned the rear of the semitrailer. In the meantime, the pickup truck had also run off the road. The pickup truck's fuel line ruptured, resulting in the truck being destroyed by fire.

As a result of the accident, the truckdriver was killed, and the driver of the pickup truck was seriously injured. Residents of five homes in the vicinity of the accident were asked to evacuate, and the highway was closed for more than 12 hours. Damage, cleanup, and lost revenues were estimated at $155,000.

The National Transportation Safety Board determines that the probable cause of the May 1, 2001, collision and subsequent fire involving a tractor-semitrailer and a pickup truck in Ramona, Oklahoma, was the failure, for unknown reasons, of the pickup driver to control her vehicle. Contributing to the severity of the accident were the inadequate protection and shielding of the cylinders, valves, piping, and fittings and the inadequate securement of cylinders on the semitrailer.

This report discusses the following safety issues:

- The adequacy of Federal requirements for protecting cylinders that are horizontally mounted on semitrailers and the valves, piping, and fittings of the cylinders during rollover accidents.

- The adequacy of the information in the *North American Emergency Response Guidebook* about compressed hydrogen.

As a result of its investigation of this accident, the Safety Board makes recommendations to the Research and Special Programs Administration.

this page intentionally left blank

Factual Information

Synopsis

About 2:15 p.m., central daylight time,[1] on May 1, 2001, a northbound tractor, in combination with a flatbed semitrailer that had horizontally mounted cylinders filled with compressed hydrogen, which is a flammable gas, struck a northbound pickup truck that had veered in front of the tractor-semitrailer on U.S. Highway 75, 2 miles south of Ramona, Oklahoma. According to witnesses, the tractor-semitrailer then went out of control and overturned, while continuing along the highway. It went off the road to the east and traveled 300 more feet before it stopped. During the process, some of the cylinders, valves, piping, and fittings at the rear of the semitrailer were damaged and released hydrogen. The hydrogen ignited and burned the rear of the semitrailer. In the meantime, the pickup truck had also run off the road. The pickup truck's fuel line ruptured, resulting in the truck being destroyed by fire.

As a result of the accident, the driver of the tractor-semitrailer (the truckdriver) was killed, and the driver of the pickup truck was seriously injured. Residents of five homes in the vicinity of the accident were asked to evacuate, and the highway was closed for more than 12 hours. Damage, cleanup, and lost revenues were estimated at $155,000.

The Accident

The tractor-semitrailer was operated by Airgas Mid-South, headquartered in Tulsa, Oklahoma. The semitrailer carried 10 horizontally mounted cylinders (see figure 1) loaded with compressed hydrogen. Each cylinder was about 22 inches in diameter and about 37 1/2 feet long. Combined, the cylinders contained a total of approximately 141,365 cubic feet[2] of hydrogen; each cylinder was pressurized to about 2,500 pounds per square inch, gauge (psig). The cylinders were secured between two bulkheads and arranged in three rows; the first (bottom) and second rows each had four cylinders, and the third (top) row had two cylinders. The cylinders had been filled with hydrogen on April 30, 2001, at the Airgas Mid-South facility in Tulsa.

---

[1] All times in this report are central daylight time.

[2] At a temperature of 70° F and at 1 atmosphere of pressure.

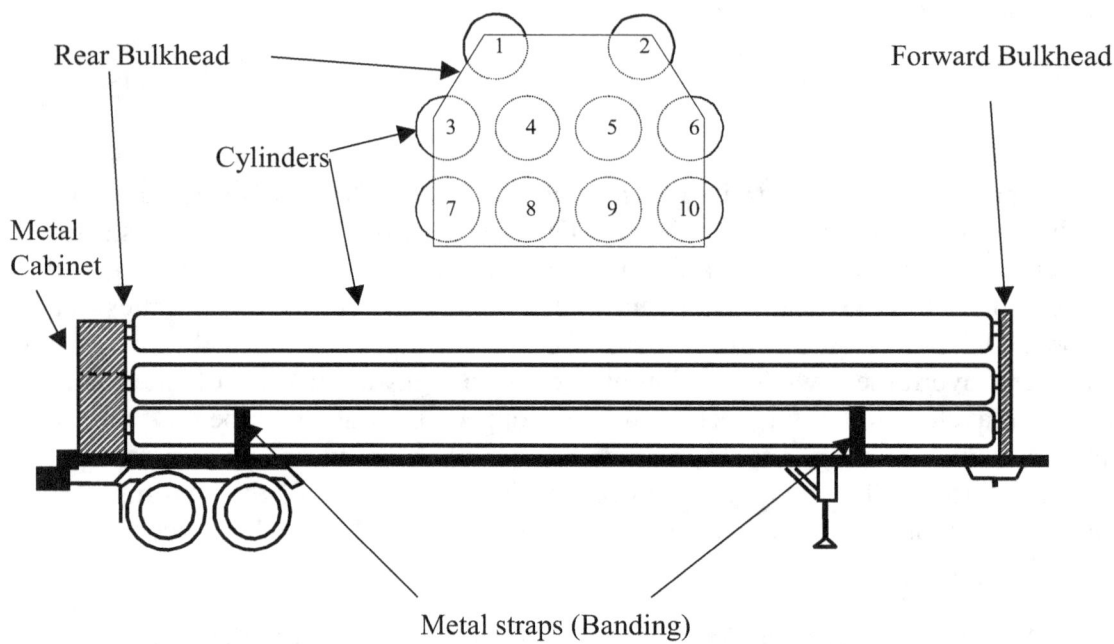

Figure 1. Mounting and cylinder configuration of the accident semitrailer.

The truckdriver departed the Airgas Mid-South terminal in Tulsa about 1:50 p.m. on May 1, 2001, destined for the Phillips Petroleum plant in Bartlesville, Oklahoma, a distance of about 45 miles. (See figure 2.) According to witnesses, minutes before the accident, the tractor-semitrailer was traveling about 65 to 70 mph[3] in the left lane of the four-lane divided highway. Ahead of the tractor in the right lane was a 1983 Chevrolet pickup truck. According to some witnesses who had followed the pickup truck for a half mile in the minutes preceding the accident, the pickup truck moved slowly from the right northbound lane to the left northbound lane of the divided highway, where it almost struck another vehicle. (One witness said that he believed the driver had fallen asleep but that when he pulled alongside the pickup truck, he noted that she [the driver] seemed to be awake.)

---

[3] The posted speed limit for U.S. Highway 75 in the vicinity of the accident was 70 mph.

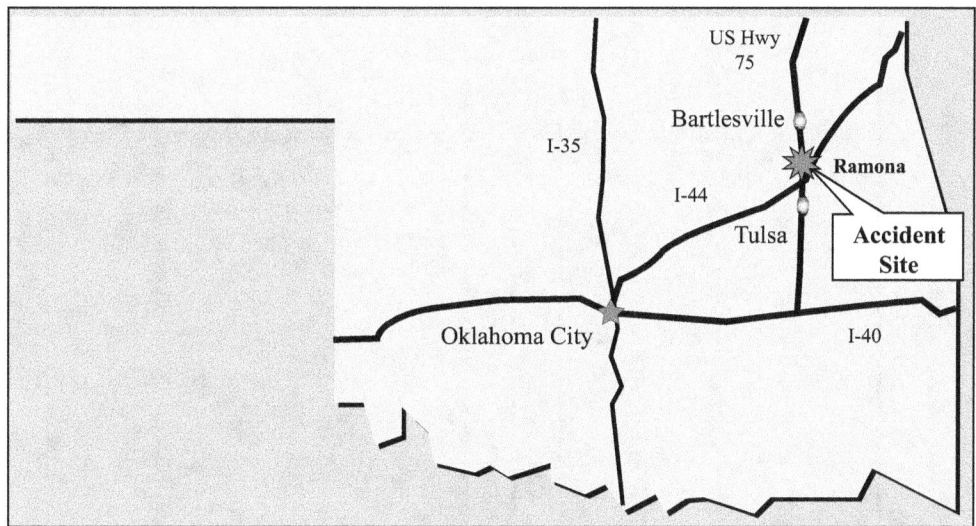

Figure 2. Accident location.

Witnesses said the pickup truck next swerved from the left lane, through the right lane, and onto the right shoulder of the highway, where it struck a one-way traffic sign. (See figure 3.) After striking the sign, the pickup truck veered back onto the highway, into the left lane and into the path of the tractor-semitrailer. The right front of the tractor struck the front left axle of the pickup truck.

Witnesses said the truckdriver appeared to be operating his vehicle normally before the accident. Witnesses stated that the truckdriver could not have avoided hitting the pickup truck.

After the collision, the pickup truck traveled about 365 feet before coming to rest against a fence and facing south. The tractor-semitrailer remained coupled and traveled about 210 feet along the roadway beyond the point where it had hit the pickup truck. The tractor-semitrailer rolled over onto its left side before leaving the roadway. The tractor-semitrailer left the road and traveled about 300 feet on the embankment before coming to rest. The tractor and the semitrailer remained coupled with the tractor upside down and the semitrailer on its right side (the tractor had rolled about 180 degrees, and the semitrailer had rolled about 270 degrees). One cylinder, cylinder 3, had separated from the semitrailer and came to rest about 53 feet away.

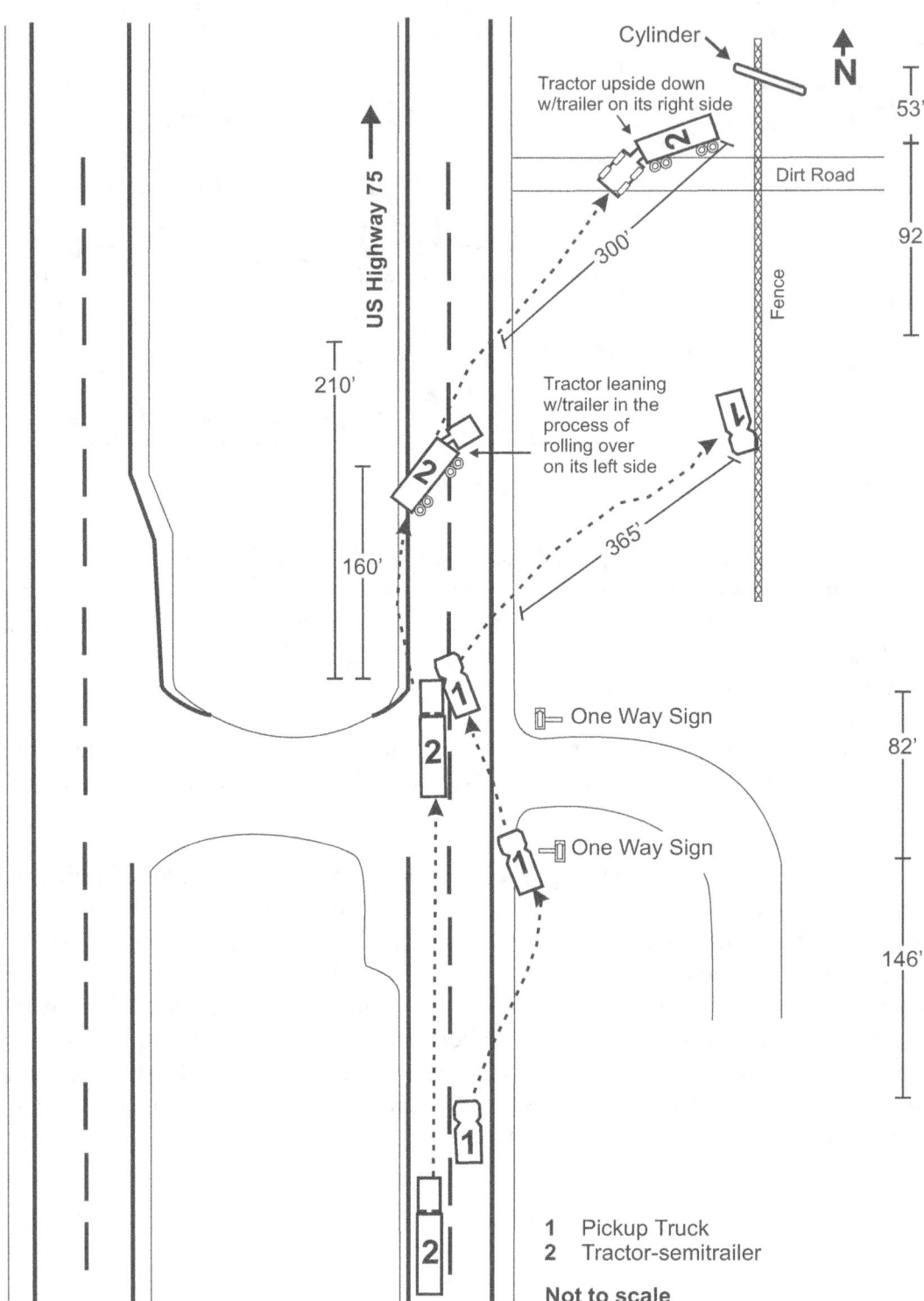

Figure 3. Accident diagram.

Emergency Response[4]

The Washington County, Oklahoma, 911 Center received numerous telephone calls at 2:15 p.m. reporting a traffic accident on U.S. 75 involving a tractor-semitrailer, explosions, and fires. Members of the Ramona volunteer fire department were dispatched to the scene. The chief of the department arrived on scene about 2:22 p.m. and assumed incident command. Before the emergency response personnel arrived, witnesses had already extricated the pickup driver from her vehicle through the passenger door. (At 3:12 p.m., an ambulance took her to St. Francis Hospital in Tulsa, Oklahoma.)

Just after the accident occurred, an Airgas Mid-South employee passed the accident and notified the company that one of its semitrailers with horizontally mounted cylinders was involved. In response, Airgas Mid-South's president, vice president, and safety director came to the scene, arriving at 3:20 p.m.

By 2:16 p.m., the Washington County Emergency Management director (the director) was en route to the accident scene and was advised by the Washington County Sheriff dispatcher that "some type of tanker was involved, possibly hydrogen." At 2:21 p.m., via radio, the director asked the Phillips Petroleum dependable accident response team (DART)[5] to come to the scene. (It arrived within 15 minutes.) At 2:22 p.m., also via radio, he told the Ramona fire department responders that the cylinders might be pressurized and asked the volunteer fire departments from surrounding areas for mutual aid.[6] Members of volunteer fire departments from the towns of Ochelata and Oglesby arrived shortly afterward.

The Ramona, Oglesby, and Ochelata fire departments initially staged their vehicles and personnel in the northbound lanes of U.S. 75, approximately 40 feet west of the overturned tractor-semitrailer and in a direct line with the forward ends of the cylinders. The vegetation that surrounded the rear of the semitrailer was burning, as were the rear tires of the semitrailer.

While the director was traveling to the scene, the Washington County assistant fire chief, who was at the fire station, read the information about hydrogen in the 1996 *North American Emergency Response Guidebook*[7] to the director by radio. At 2:25 p.m., while the director was still en route, he activated the Washington County Emergency Operations

---

[4] Refer to appendix B for a timeline of significant emergency response events.

[5] The Phillips Petroleum DART gave technical assistance in the early stages of the incident to the director. This was done in accordance with a mutual aid agreement for the DART to provide hazardous materials assistance when needed to Washington County. DART also provided two unmanned ground water monitors that were used in the latter stages of the emergency response to spray water on the cylinders.

[6] A total of nine fire departments from the surrounding communities responded to the mutual aid call.

[7] The guidebook is used to aid first responders in quickly identifying the specific or generic hazards of materials involved in hazardous materials incidents. It is published every 4 years through the cooperation of the U.S. Department of Transportation, Transport Canada, and Mexico's Secretariat of Communications and Transport. The latest edition was published in 2000.

Center and the Washington County Emergency Operations Plan.[8] At 2:30 p.m., he arrived on scene and assumed incident command from the Ramona fire department chief. By that time, the Ramona fire department had already started spraying water from several handlines on the burning vegetation and on the rear of the semitrailer.

At the same time, members of the Ochelata and Oglesby fire departments were attempting to extricate the truckdriver from the overturned and crushed tractor. He was trapped in his tractor, which was within 15 feet of the forward ends of the cylinders. According to the director, when he arrived on scene, firefighters had already made an initial medical assessment of the truckdriver. They said that they had detected a "possible weak or faint pulse" and thought the area was safe enough to extricate the driver.

At 2:35 p.m., after an initial on-scene assessment, the director decided to establish a defensive operation only and told the firefighters to spray significant amounts of water on the cylinders, "but only from a distance." Also at 2:35 p.m., the shift superintendent of the Phillips Petroleum DART arrived and told the firefighters to stay away from the ends of the cylinders. They took firefighting positions on both sides of the semitrailer and tried to avoid the ends of the cylinders. However, the emergency response vehicles that were being used to spray water were not moved away from the ends of the cylinders for another hour.

At 2:40 p.m., the effort to extricate the truckdriver was stopped, pending the delivery of additional water to continue cooling the cylinders. By 2:55 p.m., the water had arrived, and more handlines were deployed to maintain cooling streams on the cylinders. By 3:15 p.m., members of the Ochelata and Oglesby fire departments had extricated the truckdriver. Because no pulse was found, he was taken by ambulance to the medical examiner's facility in Jane Phillips Hospital, Bartlesville, Oklahoma.

In the meantime, between 2:30 and 3:11 p.m., an Oklahoma Highway Patrol officer trained in hazardous materials arrived. At 3:11 p.m., he requested, on his own initiative, the assistance of the Tulsa fire department's hazardous materials response team (the Tulsa hazardous materials team). At approximately 3:30 p.m., the first of 6 units and 17 firefighters from the Tulsa hazardous materials team arrived, as well as a pumper truck from the Bartlesville fire department.

About 3:35 p.m., the chief of the Tulsa hazardous materials team formally assumed command of the incident, after he and the director had agreed on how to evacuate personnel and equipment to a safer distance. They agreed to use the Bartlesville pumper truck and unmanned ground monitors to continue cooling operations while personnel left the area. The chief then directed the firefighters to move away from the scene. He proceeded to develop an action plan that included moving the fire suppression personnel across the highway and setting up water shuttles and unmanned monitors. He established a

---

[8] The Washington County Emergency Operations Plan is periodically certified by the Washington County commissioners, the mayor of Bartlesville, the Oklahoma emergency management planning officer, and the director. The most recent certification was on March 29, 1999.

"hot zone" perimeter of 300 to 600 feet, a "warm zone" perimeter of 600 to 900 feet, and a "cold zone" perimeter of 1,500 feet.[9] The Tulsa hazardous materials team sent a reconnaissance team in to survey the condition of the valves on the cylinders and monitored the temperature of the cylinders.

About 4:00 p.m., the Airgas Mid-South safety director offered the assistance of the Airgas AERO team[10] to the chief of the Tulsa hazardous materials team. At 4:30 p.m., the safety director activated the AERO team, and it arrived on scene about 5:15 p.m. The AERO team assisted the firefighters with cooling and monitoring operations.

About 6:00 p.m., the chief of the Tulsa hazardous materials team and the director began to discuss releasing the Tulsa fire department units because the accident scene had been stabilized. About 6:30 p.m., the Tulsa fire department units began demobilizing, and within 10 minutes, the team departed.

The director then resumed incident command. By 12:20 a.m. on May 2, all cylinders had been vented of hydrogen and the fire was out.

At approximately 12:55 a.m., the director left the scene, and the Oklahoma Highway Patrol assumed incident command. By 1:03 a.m., all fire and rescue units had left. The Oklahoma Highway Patrol and the AERO team stayed until the highway was reopened, about 6:00 a.m.

Injuries

Table 1 is based on the injury criteria of the International Civil Aviation Organization, which the Safety Board uses in accident reports for all transportation modes.

As a result of the accident, the driver of the pickup truck was seriously injured, and the truckdriver was killed. According to the Oklahoma Chief Medical Examiner, the truckdriver had sustained multiple blunt-force traumas.

---

[9] The establishment of perimeter zones helps ensure that response personnel are properly protected against the hazards present at a hazardous materials transportation accident. During this accident, the "hot zone" indicated the area where there was the greatest risk of explosion or overpressurization of the cylinders. The "warm zone" indicated the area where response personnel were staged, and the "cold zone" identified the area from which the general public and media could view the accident scene safely.

[10] The Airgas Mid-South AERO (Airgas emergency response organization) team responds to hazardous materials incidents involving Airgas Mid-South. The team is trained and equipped to handle most cryogenic and/or compressed gases.

Table 1. Injuries sustained in Ramona, Oklahoma, accident

Injuries*	Tractor-Semitrailer	Pick-up Truck	Total
Fatal	1	0	1
Serious	0	1	1
Minor	0	0	0
Total	**1**	**1**	**2**

*49 *Code of Federal Regulations* 830.2 defines fatal injury as "any injury which results in death within 30 days of the accident" and serious injury as "an injury which: (1) requires hospitalization for more than 48 hours, commencing within 7 days from the date the injury was received; (2) results in a fracture of any bone (except simple fractures of fingers, toes, or nose); (3) causes severe hemorrhages, nerve, or tendon damage; (4) involves any internal organ; or (5) involves second or third-degree burns, or any burn affecting more than 5 percent of the body surface."

Meteorological Information

The weather at the time of the accident was dry; the skies were slightly cloudy; the visibility was clear; and the temperature was near 84° F. Wind was gusting to about 17 mph from a southwesterly direction.

Damage

Pickup Truck

The pickup truck, a 1983 C-10 model Chevrolet, had sustained extensive impact and fire damage. The left side of the vehicle, including the front fender, hood, driver's door, and side of the cargo bed, was badly crushed. The left front fender had been pushed against the engine. The fire had destroyed the interior of the cab and burned away the exterior paint on the vehicle. The gas tank fuel line had separated from the fuel tank, and the tires had burned.

The steering components were damaged, and the left front wheel and tie rod were bent. The steering column and steering box had been damaged in the collision and burned in the fire. Consequently, they could not be inspected for preaccident problems. The brake lining, drums, and hydraulic lines had also been damaged enough that it was not possible to examine the brake system for pre-impact defects. Because the vehicle could not be started, the performance of the engine and related parts could not be evaluated.

Tractor-Semitrailer

Tractor-Semitrailer. The tractor, a 1999 conventional Freightliner, was badly crushed. The roof was crushed down to the dashboard. Because of the extensive crushing, the brake system on the tractor could not be fully inspected; however, the chambers and brake adjustments were checked and were within the manufacturer's readjustment limits. The data downloaded from the tractor's electronic control unit[11] showed no abnormal codes. The maintenance records for the tractor revealed only routine or expected maintenance and no known mechanical deficiencies.

The semitrailer was a 1992 Doonan standard flatbed trailer with a 10-cylinder hydrogen rack configuration welded to the bed. The semitrailer's frame rails were bent just forward of the front axle, and the torsion bar for both the front and rear axles was broken. The rear bumper was bent and twisted in various directions.

Cylinder Mounting and Equipment. Forces involved in the accident had bent both bulkheads rearward in the area where cylinder 3 (the cylinder ejected from the semitrailer) had been mounted. The resulting deformation left the two bulkheads about an inch further apart in this area than they had been designed to be. The paint on the rear bulkhead had been damaged by the fire. The bands (metal straps) used to secure the bottom row of cylinders (cylinders 7 through 10) were still intact.

The cabinet, which enclosed the valves, piping, and fittings for the cylinders, had separated from the rear bulkhead and was crushed inward from the passenger side of the semitrailer toward the driver's side. (See figure 4a. Figure 4b shows the rear view of a similar vehicle configuration without damage.) The master valve and the shutoff valves, piping, and fittings for eight of the cylinders had been sheared off. The valves for the other two cylinders were intact but broken away from their piping. Both valves were bent in a direction consistent with the application of a force applied from the right side of the trailer.

Cylinders. The paint on the rear end of most of the cylinders (cylinders 1, 2, 4, 7, 8, and 9) was sooted and discolored. The sooted area was directly over the rear tires, which were shredded and burned. The paint on the outboard side of each outside cylinder (cylinders 1, 2, 3, 6, 7, and 10) was, to varying degrees, scraped along the length of each cylinder. (See figure 5.) These cylinders extended beyond the envelope formed by the bulkheads. (Refer to figure 1.) Cylinder 1 also had scraped paint that extended around its circumference in a 2- to 3-foot-wide strip near the middle of the cylinder. The unpainted threads at each end of cylinder 1 were also visible. (The unpainted threads cannot be seen when a cylinder is fully screwed into flanges bolted to the front and rear bulkheads. The flanges secure the front and rear of the cylinder to the front and rear bulkheads of the semitrailer.) (See figure 6.)

---

[11] This is a semiconductor unit for controlling ignition timing and other parameters in an engine management system.

Remains of Metal Cabinet

Damage to piping
and valves

Twisted and bent bumper

Figure 4a. Cabinet after accident; 8 of the
10 valves have been sheared off.

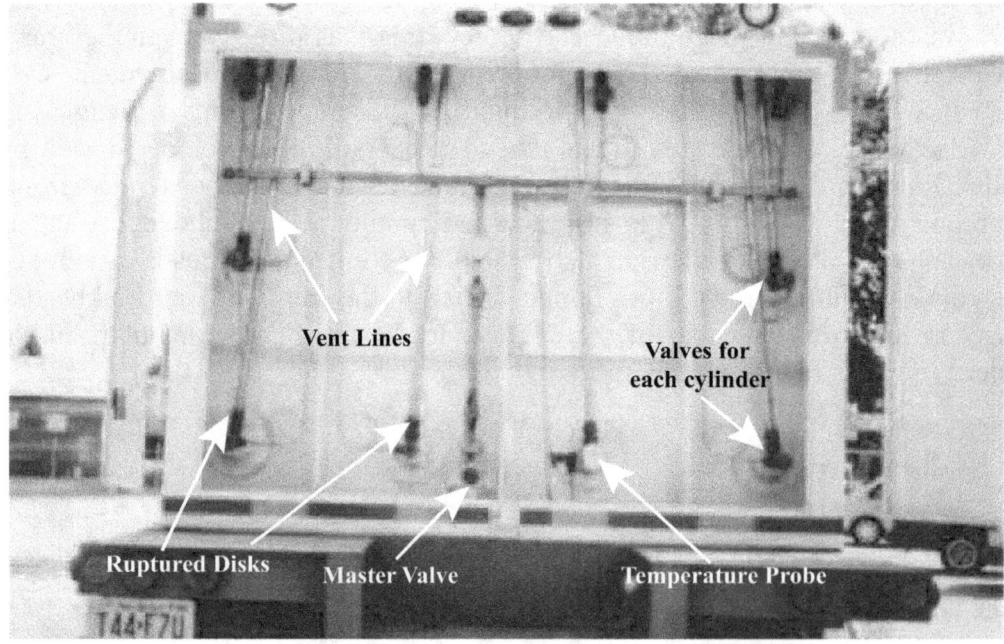

Vent Lines

Valves for
each cylinder

Ruptured Disks   Master Valve   Temperature Probe

Figure 4b. Rear view of a typical semitrailer's cabinet without damage.

Figure 5. Scrapes on the outboard sides of cylinders on the accident semitrailer.

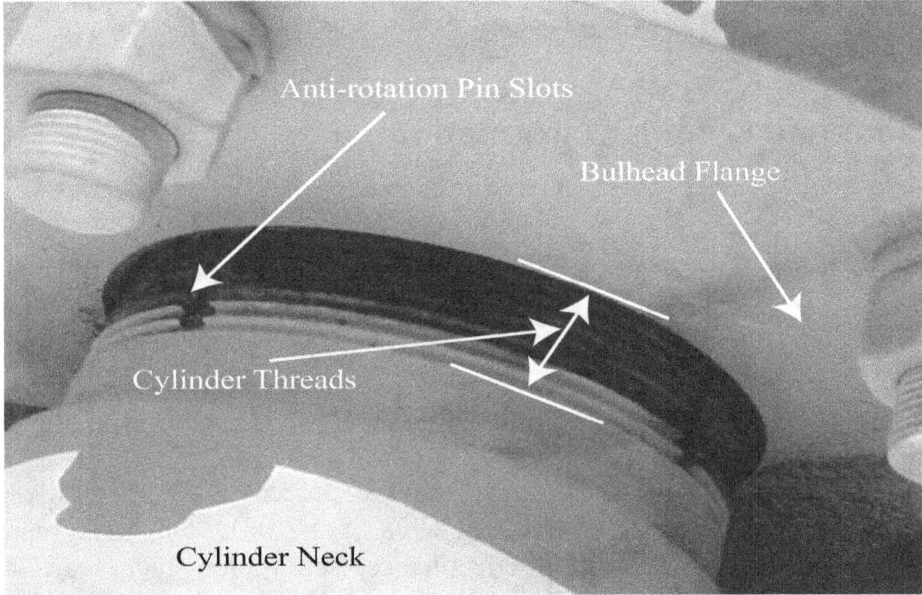

Figure 6. Unpainted threads of cylinder 1 next to the bulkhead flange.

Cylinder 3, which had been ejected from the vehicle, also had circumferential abrasions. The front neck/shoulder of the cylinder was fractured (see figure 7). The fracture features and deformation were consistent with an overstress fracture. The threads on the rear neck of the cylinder and the threads on the rear flange were also damaged, with off-axis rotational scoring.

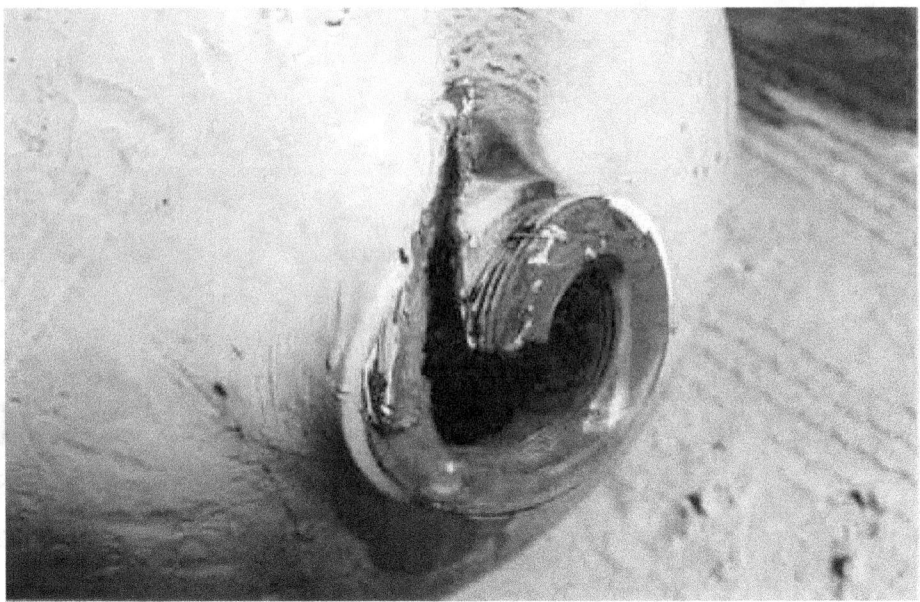

Figure 7. Cylinder 3's forward end, which has an overstress fracture on its shoulder and a spiral crack from the fracture surface into the head of cylinder.

Semitrailer and Cylinder Information

Semitrailer

The accident semitrailer had been manufactured in October 1991 by Doonan Trailer Corporation of Great Bend, Kansas. Doonan had also mounted the cylinders on the semitrailer. The accident semitrailer was one of seven semitrailers with horizontally mounted cylinders that Doonan manufactured between the late 1980s and early 1990s for Brown Welding[12] of Salina, Kansas. Doonan no longer produces this type of semitrailer.

Today, four manufacturers[13] produce about 80 percent of the semitrailers that have horizontally mounted cylinders. Between 1998 and 2001, the four manufacturers produced, on average, a total of about 240 of these semitrailers each year;[14] the largest manufacturer produced 120 annually, and the smallest produced 18. Safety Board investigators tried to ascertain the number of these semitrailers currently in operation; however, no government agency or industry organization could provide an accurate accounting of the operational fleet.

---

[12] Airgas purchased Brown Welding in 1999.

[13] FIBA Technologies, Inc.; City Machine and Welding; Weldship Corporation; and Western Sales and Testing.

[14] The number includes new semitrailers with used cylinders that have been requalified under 49 CFR 173.34 and new cylinders manufactured overseas.

Cylinders

The 10 cylinders on the accident semitrailer were U.S. Department of Transportation (DOT) specification 3AAX cylinders and were manufactured in 1991 by CP Industries (CPI) of McKeesport, Pennsylvania. All 10 cylinders were certified acceptable for transporting compressed gases following a postmanufacture inspection conducted in July 1991 by the Robert W. Hunt Company of Lombard, Illinois, as required by the DOT's hazardous materials regulations.

CPI manufactures approximately 600 DOT specification 3AAX cylinders annually. CPI also manufactures about 800 DOT specification 3T cylinders annually. CPI is the only known domestic manufacturer of specification 3AAX and 3T cylinders. Both types of cylinders are used to transport virtually the same products and are subject to the same design and operating requirements under the hazardous materials regulations. The specification 3AAX cylinder is authorized to transport the following non-liquefied compressed gases: air, argon, boron trifluoride, carbon monoxide, ethane, ethylene, helium, hydrogen, methane, neon, nitrogen, and oxygen. The specification 3T cylinder is authorized to transport all of those gases except hydrogen. Both specifications must be horizontally mounted on motor vehicles or in an ISO (International Organization of Standardization) framework.[15] The specification 3AAX and 3T cylinders have different design requirements relating to materials of construction, thickness of the cylinder wall, and maximum allowable pressure ratings.

Horizontally mounted cylinders are typically about 37 feet long and 22 inches in diameter. A single semitrailer carries between 7 and 12 cylinders. Each cylinder on the accident vehicle had a water capacity of about 87 cubic feet and a minimum wall thickness of 0.536 inch.

Both necks of each cylinder on the accident semitrailer had inside and outside threads. The outside threads were screwed into six-bolt flanges that secured the cylinder to the bulkheads. Anti-rotation pins were inserted between the flange and the outside threads of the cylinder. (See figure 8.) Adaptors were screwed into the inside threads of each neck of each cylinder. The front-end adaptor held a rupture disk,[16] whereas the rear-end adapter held a rupture disk and the shutoff valve for the cylinder. The valves, piping, and fittings at the rear of the semitrailer allowed the contents of all 10 cylinders to be loaded or unloaded simultaneously or in a series of one or more. The valves and piping from each cylinder joined at a master shutoff valve.[17] Piping used to release pressure and vent each cylinder was installed on the front and rear of the semitrailer.

---

[15] An ISO framework, as it applies to horizontally mounted DOT specification 3AAX and 3T cylinders, provides a box-like structure around a fixed number of cylinders. This configuration and design provides structural integrity and support to the cylinders it transports.

[16] Rupture disks are also known as safety disks. They are frangible safety devices that burst and release built-up pressure so the cylinder would not burst if overpressurized. The rupture disks on the cylinders had a burst range of 3,700 to 4,000 psig.

[17] The piping from each cylinder had a diameter of 3/8 inch, and the vent piping had an outer diameter of 1 1/8 inches.

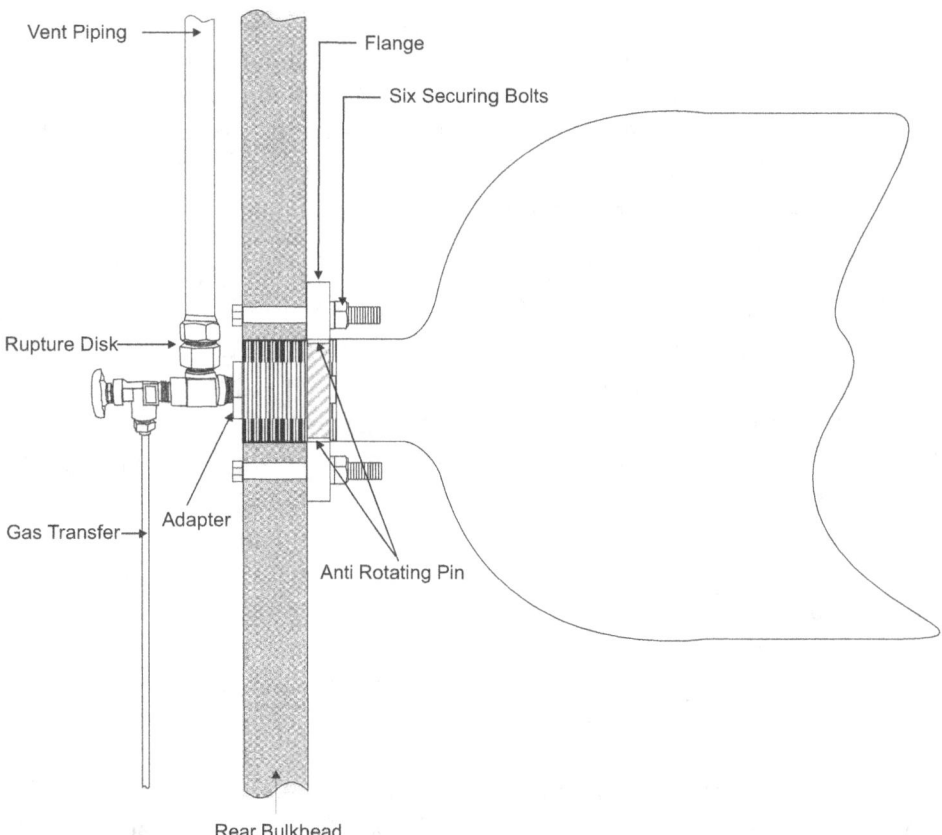

Figure 8. Diagram of the rear end of a cylinder, showing how it is secured into a flange, which is then secured with bolts to the bulkhead. The valve is screwed into an adapter, which is inserted into the end of the cylinder.

Hazardous Material Information

Hydrogen is a colorless, odorless, highly flammable gas. Hydrogen is a serious fire hazard when accidentally released. Hydrogen ignites readily in air at any ambient temperature. It burns with an invisible flame; and because it is less dense than air, it rises when it is released to the atmosphere. Hydrogen has a lower explosive limit of 4 percent and an upper explosive limit of 75 percent.[18] (For comparison, natural gas has a lower explosive limit of 3.8 percent and an upper explosive limit of 17 percent.)

---

[18] Lower and upper explosive limits refer to the concentration of a gas in air that will support combustion.

Personnel Information

Tractor-Semitrailer Driver

The truckdriver was a 46-year-old male employee of Airgas Mid-South. He had a Kansas commercial driver's license with a hazardous materials endorsement that had an expiration date of November 17, 2003. The Oklahoma Highway Patrol had no record of violations by the driver.

According to medical records obtained from his personal physician, the truckdriver was being treated for multiple medical problems, including diabetes (treated with Actos [pioglitazone], Glucophage [metformin], glyburide, and insulin), chronic neck pain for which he had been seeking disability (treated with disk surgery and spinal fusion), and anxiety (treated with Xanax [alprazolam]). His personal physician first prescribed insulin for him on October 16, 2000, after oral medications alone failed to control his blood sugar. Insulin and the oral medications that he used may cause side effects, particularly abnormally low blood sugar (hypoglycemia), which can impair performance. Common side effects of Xanax include drowsiness and light-headedness; the medication has also been shown to impair the performance of many cognitive and psychomotor tasks.

A note signed by the driver's personal physician on November 8, 2000, in his medical records stated that the driver's "diabetes has been under great control, and he has had no problems" on the prescribed oral medications and insulin. An additional note in his medical records, which was dated January 31, 2001, indicated that a prescription for insulin had been called in to the driver's pharmacy and that the driver should "stay on it."

On November 28, 2000, about 5 months before the accident, the driver had had his most recent required physical examination.[19] According to the medical examiner, the driver told her that he had diabetes and that he was taking three prescribed oral medicines for diabetes: Actos, Glucophage, and glyburide. He did not tell her that he was taking insulin for his diabetes. According to DOT motor carrier regulations for commercial drivers found at 49 CFR 391.41(b)(3), a driver with an "established medical history or clinical diagnosis of diabetes mellitus currently requiring insulin for control" cannot be medically certified.

The medical examiner checked the box on the certification form that indicated that the driver suffered "permanent defect from illness, disease or injury." She entered the following on the certification form: "Fusion of C4-C5," "Type 2 Diabetes,"[20] the names of the three diabetes medications that the driver said he was taking, and "healthy male." She signed the form that certified the driver was medically qualified to operate a commercial motor vehicle.

---

[19] Pursuant to 49 CFR 391.45, all operators of commercial motor vehicles must be medically examined and certified at least every 2 years.

[20] Non-insulin-dependent diabetes mellitus; diabetes mellitus -- Type II.

A comprehensive toxicology evaluation of the blood specimen obtained during the driver's autopsy was conducted by the Civil Aeromedical Institute (CAMI), Oklahoma City, Oklahoma, and was negative for tested drugs, including alcohol, cocaine, amphetamines, marijuana, phencyclidine, opiates, benzodiazepines, barbiturates, antidepressants, antihistamines, meprobamate, methaqulaone, and nicotine at established cutoff values.[21] Hemoglobin A1c in the specimen, an indication of blood sugar control, was minimally elevated.[22]

Pickup Driver

The pickup driver was 69 years old and had a valid Oklahoma motor vehicle license with an expiration date of August 2004. Her driving record did not include any violations, and her license had no operating restrictions.

Her medical records indicated that she was taking prescription medications for several problems, including diabetes (treated with Glucophage XR [metformin extended-release] and Glucotrol [glipizide]), hypertension (treated with Tarka [verapamil/trandolapril]), and anxiety (treated with Triavil [amitriptyline/perphenazine]). Glucophage XR and Glucotrol are oral medications for diabetes. Side effects may include abnormally low blood sugar (hypoglycemia), which can impair performance. Tarka is used to treat high blood pressure. Side effects may include abnormally low blood pressure. Triavil is typically used by patients with moderate to severe anxiety and depression. Amitriptyline, a component of Triavil, has sedative effects. Perphenazine, the other component of Triavil, is an anti-psychotic medication. Triavil may cause tardive dyskinesia, a condition marked by involuntary muscle spasms and twitches in the face and body.

According to the driver's medical records, her blood pressure and blood sugar were normal on medications. A comprehensive toxicology evaluation of a blood specimen obtained during her initial medical evaluation after the accident was conducted by the CAMI and was negative for the following drugs: alcohol, cocaine, amphetamines, marijuana, phencyclidine, opiates, benzodiazepines, barbiturates, antidepressants, antihistamines, meprobamate, methaqulaone, and nicotine at established cutoff values. Hemoglobin A1c in the specimen, an indication of blood sugar control, was normal.

On May 5, 2001, Safety Board investigators briefly interviewed the driver of the pickup truck in her hospital room. She was asked to describe the events that led up to the accident. She stated that she would not have lost control of her pickup truck if the tractor-trailer had not hit her first. Due to her medical condition and the length of time required for additional questioning, the investigators later followed up with questions about the medications she took and her activities during the 3 to 4 days before the

---

[21] These values are the threshold values in μg/ml used to report positive results. Values below these concentrations are normally reported as not detected.

[22] The level reported was 6.8 percent; at the CAMI laboratory, "levels above 6% are considered abnormal."

accident. She responded by providing a list of her medications and the name of her physician, but she did not recall her activities for the 3 to 4 days before the accident.

Roadway Information

U.S. Highway 75, in the vicinity of the collision, was a divided four-lane concrete roadway running north-south that had a posted speed limit of 70 mph. The paved portion of the northbound roadway was about 39 feet wide, which included an 11-foot-wide outside (right) shoulder, two 12-foot-wide traffic lanes, and a 4-foot-wide inside (left) shoulder. East of the northbound outside shoulder was an earthen embankment with an approximate slope of 40 percent and an 80-foot easement to the fence line.

The traffic lanes were divided by standard pavement markings, painted white stripes that were approximately 12 feet long and 30 feet apart. Painted edge lines separated the traffic lines from the shoulders. A solid white edge line (fog line) ran along the outside shoulder, and a solid yellow edge line ran along the inside shoulder. The northbound and southbound portions of the highway were separated by a depressed earthen median. The part of the median that was directly adjacent to the accident location was about 56 feet wide.

Motor Carrier Information

Airgas is a national distributor of compressed gases operating at 700 locations throughout the United States. Airgas Mid-South, which is 1 of 14 regional divisions of Airgas, employs about 550 people and operates 45 retail/wholesale outlets,[23] 8 cylinder-filling plants, and 1 ultrasound testing facility in an area that includes western Missouri, northern Texas, Arkansas, Louisiana, Oklahoma, Kansas, western Mississippi, and Tennessee.

The company buys hydrogen and other flammable and non-flammable gases from various suppliers and transports the gases to customers by using either semitrailers with horizontally mounted cylinders or other vehicles that carry portable cylinders. The customers may be industrial, medical, manufacturing, or maintenance facilities. Because Airgas Mid-South is one of a few regional companies to deliver hydrogen, it delivers hydrogen as far northeast as Michigan and as far west as Phoenix, Arizona. Airgas Mid-South has 128 regular hydrogen customers and delivers approximately 207 hydrogen shipments each month by semitrailers with horizontally mounted cylinders. Airgas Mid-South owns 14 semitrailers with horizontally mounted cylinders and leases an additional 10.

---

[23] The company sells industrial, medical, and specialty gases; welding equipment; and safety equipment and supplies at these locations.

Federal Regulations

Protection of Valves, Piping, Fittings, and Cylinders

Under the hazardous materials regulations (49 CFR 173.301(l)(2)), specification 3AAX and 3T cylinders when transported on a motor vehicle or in an ISO framework must be fixed at one end of the vehicle or framework with "provision" for thermal expansion at the opposite end attachment. The "protective structure" for valves and safety relief devices for the cylinders must be "sufficiently strong to withstand a force equal to twice the weight involved with a safety factor of four, based on the ultimate strength of the material used." The regulations do not define the term "weight involved" or explicitly state whether the safety factor is based on the weight of the vehicle or on the ultimate strength of the material of the protective structure.

During a postaccident interview with Safety Board investigators on July 12, 2001, the acting director for the Research and Special Programs Administration's (RSPA's) Office of Hazardous Materials Technology defined "weight involved" as the weight of the entire vehicle applied in a single direction from the rear and said that the term does not include side or vertical forces. He said that the only valve protection requirements applicable to horizontally mounted cylinders on semitrailers are those in the section cited above and that the section has not been revised since 1964. He also stated that the cabinet and rear bumper on a semitrailer provide additional protection and security for the valves and fittings.

The hazardous materials regulations do not require that horizontally mounted cylinders also be protected or shielded from the impact damage resulting from a rollover of the semitrailer. Further, the regulations require only that horizontally mounted cylinders be secured at both ends of a semitrailer and do not require additional measures to reduce the likelihood of a cylinder being ejected during a rollover accident.

Industry Practices and Interpretations

The four manufacturers of semitrailers with horizontally mounted cylinders and CPI said they knew about the general requirement (49 CFR 173.301) that valves be protected. One manufacturer stated that the cabinets are designed to protect the valves from being sheared or broken off. According to another manufacturer, the extension of the trailer past the rear tires serves as the valve protection, and the cabinet is intended simply for shielding the valves from inclement weather.

The four manufacturers stated that although they are not required to band cylinders on semitrailers, all four use banding as an extra precaution to keep the cylinders on the semitrailer. Three of the four manufacturers stated that they use secondary securing on at least some of the cylinders on their semitrailers. The fourth manufacturer stated that it bands all cylinders on its semitrailers. One manufacturer stated that because its bulkheads are so strong, it bands only the bottom row of cylinders.

Emergency Response Preparedness

Postaccident Assessment

The Washington County Emergency Management director stated during postaccident interviews with Safety Board investigators that when he arrived on scene, he was troubled by how close the emergency response vehicles were to the burning semitrailer. He said that the majority of responding vehicles should have been further away from the tractor and semitrailer. The director believed that the responding vehicles were positioned in a way that limited their escape route (instead of being backed in, they had been pulled in front first). He stated that "based on the training level of some of the initial responders, they should have known better" because some of the firefighters had been through hazardous materials training. The director said he considered the risks of breaking down the firefighting equipment and moving all personnel and equipment back to a safer location. He felt that the scene appeared to be stable and, therefore, decided that the risks involved in breaking down and repositioning personnel and equipment seemed to be greater than the risks involved in keeping them in place and maintaining the stability of the scene. He was concerned that the scene could deteriorate very quickly if the responders broke down and restaged their equipment. He requested that some of the responding vehicles (an ambulance, an Oklahoma Highway Patrol car, and his Suburban) not involved with the firefighting efforts be repositioned to the north. His goal was to reach the isolation distance given in the guidebook (1 mile in all directions) as quickly as possible.

Training

According to the director, in Oklahoma, neither State nor county government imposes formal training requirements on volunteer fire departments. The level of training for the volunteers ranges from none to level II firefighter, which is the most advanced course in firefighting offered by Oklahoma State University. A level II firefighter must be able to (1) determine the need for command, (2) organize and coordinate an incident management system until command is transferred, and (3) function within a single role in the incident management system. Such a firefighter is trained in communications, fire ground operations, rescue operations, prevention, preparedness, and maintenance. According to the director, most Oklahoma fire departments are beginning to require volunteer firefighters to complete a course on volunteer firefighting practices that covers limited interior structural (building) firefighting techniques and is also offered by the university.

A majority of volunteer firefighters in Washington County have had hazardous materials awareness training, which teaches firefighters how to recognize the presence of hazardous materials and, if possible, to identify them. Some of the firefighters have also had hazardous materials operations training, which teaches basic hazard and risk assessment techniques and basic hazardous materials terms. A select few firefighters are qualified as hazardous materials technicians; they are trained to classify, identify, and verify known and unknown materials by using field survey instruments and to understand hazard and risk assessment techniques. According to the director, "a majority of the departments and personnel involved in [this accident] have trained pretty extensively in comparison to other volunteer departments in the State." According to his training records, the director had completed the following relevant courses: Basic Incident Command System, Evacuation Operations, Hazardous Materials Awareness Instructor, On-Scene Incident Commander, and Fire Service Instructor I.

On March 17, 2001, about 2 months before the accident, Washington County Emergency Management held an incident command drill. It was a 4-hour exercise that involved a controlled burn of a meadow behind Phillips Petroleum. The participating organizations and agencies included the Phillips Petroleum DART, county fire departments, and other county emergency management officials. The purpose of the exercise was to practice the implementation of a unified command structure and the use of the county emergency operations center.

Postaccident Actions

According to the director, after the accident, Washington County established a hazardous materials response team. As of May 7, 2002, this team became active on a limited basis for immediate emergencies. The team is made up of volunteer firefighters from six area departments. The team's base is the Washington County Emergency Operations Center in Bartlesville, Oklahoma. The team has purchased a trailer and a limited amount of equipment. As a direct result of the Ramona accident and of the terrorist attacks of September 11, 2001, the county received a $35,000 grant to acquire more equipment. According to the director, the team will coordinate training on its own equipment with Oklahoma State University. All the team members have received hazardous materials operations training, which includes learning how to recognize hazardous materials; and an unspecified number of team members have received advanced hazardous materials training.

Since May 1, 2001, the county has also performed one community-wide drill, in fall 2001, with the participation of Phillips Petroleum. The scenario involved a chemical spill and fire resulting in simulated injuries requiring hospitalization and simulated fatalities. In addition, each volunteer fire department in the county is holding drills within its department. The Washington County fire department, for example, has had drills on the use of personal protective equipment. Since November 2001, the county has required each fire department to complete 3 hours of training in firefighting techniques every month.

North American Emergency Response Guidebook

According to the *North American Emergency Response Guidebook*, the book is to be used by first responders during the initial phase of a hazardous materials incident. The book states that it "is primarily a guide to aid first responders in quickly identifying the specific or generic hazards of the material(s) involved in the incident and protecting themselves and the general public during the initial response phase of the incident." The book explains that it should not be considered a substitute for emergency response training, knowledge, or sound judgment.

The book has 61 different hazardous materials guides that explain initial response measures for all of the designated classifications of hazardous materials, such as flammable gases, corrosive materials, and poisons. Each guide provides information about the potential of fire and/or explosion and about health hazards. Each guide also recommends appropriate levels of protective equipment for emergency responders, isolation and evacuation distances, and firefighting and first aid measures.

The guide for hydrogen also applies to other flammable gases, such as liquefied petroleum gas, and refrigerated liquids, such as compressed trifloroethane (a refrigerant). The guide for hydrogen recommends that firefighters fight the fires "from a maximum distance" or use unmanned hose holders or monitors. The guide does not list some of the properties that are unique to hydrogen; it does not explain that hydrogen burns with an invisible flame front and is lighter than air. Other gases to which the guide applies burn with a visible flame front or are nonflammable, are heavier than air, and sink to the ground.

Analysis

This analysis is presented in three main parts. In the first part, the Safety Board identifies factors that can be readily excluded as causal or contributory to the accident. In the second part, the Board analyzes the cause of the accident. In the final part, the Board discusses the safety issues arising from the investigation:

- The adequacy of Federal requirements for protecting cylinders that are horizontally mounted on semitrailers and the valves, piping, and fittings of the cylinders during rollover accidents.

- The adequacy of the information in the *North American Emergency Response Guidebook* about compressed hydrogen.

Exclusions

The highway in the area of the accident was generally level and straight. The visibility at the time of the accident was excellent, and the pavement was dry. The postaccident review of mechanical and maintenance records for the tractor and the semitrailer indicated that both had been properly maintained. The postaccident inspection of the pickup truck, the tractor, and the semitrailer was limited by the damage the accident had inflicted on them. To the extent that inspection was possible, however, no mechanical problems were found. When the driver of the pickup truck was interviewed, she did not note any mechanical problems with her vehicle. According to the toxicological tests of both drivers, neither had used alcohol or illicit drugs. The Safety Board, therefore, concludes that the following factors can be excluded as causal or contributory to the accident: highway conditions, weather, the mechanical condition of the vehicles, and impairment from alcohol and illegal drugs.

The Accident

Witnesses who had followed the pickup truck for about a half mile before the accident occurred said that the pickup truck driver's driving was erratic and that in the last moments before the accident, the pickup truck left the roadway briefly and struck a highway sign before veering back onto the highway and directly into the path of the tractor-semitrailer. According to those witnesses who saw the tractor-semitrailer moments before the accident, the truckdriver was operating his vehicle normally and had no opportunity to avoid the pickup truck. As a result of the collision, he lost control of his vehicle, which rolled over and began to release hydrogen, which subsequently ignited. The assertions of the pickup driver that she would not have lost control of her vehicle if the tractor-trailer had not hit her first are not supported by witness accounts or physical evidence (damage to pickup truck, tractor, and one-way sign).

Driver Performance

Pickup Driver

Nothing in the driving history of the pickup driver explains her erratic driving just before the accident. However, her medical conditions or medications may conceivably have played a role. Her diabetes and high blood pressure seemed to be well managed, and the medications she took for those conditions are unlikely to impair the user. However, her anxiety and/or the medication prescribed to treat it could have affected her ability to control the vehicle. Her anxiety medication has been reported to cause movement disorders.

Because the volume of the toxicological specimen of the pickup driver's blood supplied to the CAMI was limited, the initial screening probably would not have detected the various drugs she was prescribed. Further, the prescription drugs that she was taking, metformin, glipizide, and perphenazine, are not detectable by the routine screening procedures used at the CAMI; and the prescribed doses of verapamil and amitriptyline she took would be expected to be lower than the laboratory's cutoff values.

The Safety Board has previously addressed the issue of medication use by transportation operators (both commercial and private) and has issued recommendations to the DOT, the modal administrations, and the U.S. Food and Drug Administration (FDA) regarding the need for additional guidance for transportation operators about the potential effects of prescription and over-the-counter medications.[24] In addition, the Safety Board and the FDA co-sponsored a public meeting called "Transportation Safety and Potentially Sedating or Impairing Medications" on November 14 through 15, 2001.

The pickup driver's driving may have been erratic because she was impaired by a medical condition, a medication, or fatigue. However, the Safety Board concludes that because the pickup truck driver was unable to recall information about her activities, including her use of medications and sleep in the 96 hours before the accident, the Safety Board cannot determine with any confidence why her driving was erratic or why she lost control of her vehicle.

---

[24] On January 13, 2000, the Safety Board issued Safety Recommendations I-00-1 through -5, A-00-4 through -6, H-00-12 through -15, R-00-2 through -8, and M-00-1 through -4. Safety Board staff members have met with representatives of the DOT modal administrations to discuss implementation of these recommendations.

Truckdriver

At his last required physical examination, when he was questioned about his prescription medications and illnesses by the medical examiner, the truckdriver failed to disclose that he took insulin to control his diabetes or that he was on medication for anxiety. There is no guidance in the DOT motor carrier regulations about medically certifying commercial drivers who are being treated for anxiety or chronic pain,[25] but insulin-dependent diabetes is specified as a medical disqualification. The Safety Board explored the issue of the inappropriate medical certification of commercial drivers in its report on a May 9, 1999, bus accident in New Orleans, Louisiana.[26] In the report, the Board issued 10 recommendations to the Federal Motor Carrier Safety Administration (FMCSA) and the American Association of Motor Vehicle Administrators, urging them to develop comprehensive medical oversight programs for commercial drivers. (See appendix C.) The Safety Board is currently evaluating the FMCSA's response to Safety Recommendations H-01-17 through -25. The American Association of Motor Vehicle Administrators has not yet responded to Safety Recommendation H-01-26.

The prescription drugs that the truckdriver was taking, pioglitazone, metformin, alprazolam, and glyburide, are not detectable by the routine screening procedures used at the CAMI. He was also taking insulin, which is a naturally occurring substance in the body and is not tested for in routine toxicological testing.

Although the combination of medications that the truckdriver was taking for diabetes and the medication he was taking for anxiety could have reduced his ability to operate a commercial motor vehicle safely, witnesses said he appeared to be operating his vehicle normally before the accident. Also according to witnesses, he could not have avoided hitting the pickup truck. No physical evidence, such as mechanical difficulties or road conditions, contradicts the witness accounts. The Safety Board, therefore, concludes that the truckdriver's operation of the tractor-semitrailer did not cause or contribute to the accident.

Performance of Cylinders, Valves, Piping, and Fittings

Damage to Valves, Piping, and Fittings

During the accident and rollover, nearly all of the shutoff valves, piping, and fittings for the 10 cylinders were destroyed by the impact with the roadway and/or terrain. Eight of the 10 shutoff valves were sheared off, resulting in the release and ignition of the hydrogen. The master valve was also destroyed by impact damage. The metal cabinet

---

[25] Title 49 CFR 391.41 does, however, note disqualification for "...rheumatic, arthritic, orthopedic, muscular, neuromuscular, or vascular disease which interferes with his/her ability to control and operate a commercial motor vehicle safely" and for "...mental, nervous, organic, or functional disease or psychiatric disorder likely to interfere with his/her ability to drive a commercial motor vehicle safely."

[26] National Transportation Safety Board, *Motorcoach Run-Off-The-Road Accident, New Orleans, Louisiana, May 9, 1999*, Highway Accident Report NTSB/HAR-01/01 (Washington, D.C: NTSB, 2001).

enclosing the valves, piping, and fittings was also heavily damaged during the accident and rollover; the cabinet was not sufficiently sturdy to shield and protect the valves and fittings from impact damage. The extended rear bumper on the semitrailer, which some manufacturers described as protection for the valves and fittings, was also damaged. More importantly, on the basis of the damage to the metal cabinet, the impact forces came from the right side of the semitrailer when it swung around and impacted the roadway and terrain during its 270-degree rollover. The only protection an extended rear bumper can provide is from a force applied directly from the rear.

The configuration of the valves and fittings on the rear of the trailer also increased their vulnerability to impact damage. The shutoff valves were not positioned at the necks of the cylinders or recessed in the necks of the cylinders. The piping had limited strength, given its small diameter and thin walls. The inadequacy of the protection and shielding of the valves and fittings was critical in allowing the release of the compressed gas. Therefore, the Safety Board concludes that because the valves, piping, and fittings at the rear of the semitrailer were not adequately protected and shielded from the impact caused by the rollover of the semitrailer, 8 of the 10 shutoff valves were sheared off, which resulted in the release and ignition of the hydrogen

Fracture and Ejection of Cylinder 3

The metallurgical examination of the fracture of cylinder 3 showed that the fracture was caused by overstress that resulted from forces acting on the cylinder during the rollover. The fracture was consistent with the impact of the front of the cylinder with the roadway and/or terrain. Because the outermost cylinders extended beyond the top and side edges of the bulkheads, the cylinders were exposed and vulnerable to absorbing the initial impact during a rollover. (See figure 1.)

The rear neck of the cylinder was also subjected to motions and forces that resulted in the cylinder rotating loose from the flange, as demonstrated by the off-axis rotational scoring observed in the interior threads. A similar rotation of cylinder 1 was demonstrated by the exposed and unpainted threads on its front and rear necks. Therefore, the Safety Board concludes cylinder 3 absorbed the initial impact with the roadway or terrain, resulting in the fracture of its front end and the ejection of the cylinder from the semitrailer.

Standards for Cylinders Horizontally Mounted on Semitrailers

Under the hazardous materials regulations, protection for valves and rupture disks for DOT specification cylinders horizontally mounted on semitrailers must be designed to "withstand a force equal to twice the weight involved with a safety factor of four, based on the ultimate strength of the material used." Although RSPA has not issued a formal written interpretation of this requirement, RSPA staff has advised the Safety Board that the requirement applies only to a rear-end strike to the semitrailer. An extended rear bumper may offer protection from rear strikes, but it affords no protection from the side and top forces that are typically encountered in a rollover accident.

The regulations do not specify whether the valves, piping, and fittings must be enclosed and shielded by a protective structure; and the regulations do not explain whether other options for protection, such as recessed valves and fittings, are acceptable. This lack of specificity in the regulations led one manufacturer to believe that the cabinet enclosing the valves, piping, and fittings is designed to protect them from being sheared or broken off. A second manufacturer believes that an extended rear bumper fulfills the requirements to protect valves and fittings. Another manufacturer stated that the cabinet must be constructed to comply with the requirements that apply to portable gas cylinders. There is no clear and concise requirement in the hazardous materials regulations that addresses the protection of valves and fittings from forces in a rollover accident. The absence of RSPA guidance has created differing perceptions within the industry about what is actually required for protecting valves, piping, and fittings on a semitrailer with horizontally mounted cylinders.

Because of the ambiguities of the existing regulations and the various interpretations among the manufacturers of the vehicles, the Safety Board concludes that the hazardous materials regulations do not provide sufficient and clear requirements for protecting cylinders and valves, piping, and fittings of cylinders that are horizontally mounted on semitrailers. Consequently, the hazardous materials regulations must address the multidirectional forces that these cylinders, valves, piping, and fittings may experience in a rollover accident. Therefore, the Safety Board believes that RSPA should modify 49 CFR 173.301 to clearly require that valves, piping, and fittings for cylinders that are horizontally mounted and used to transport hazardous materials are protected from multidirectional forces that are likely to occur during accidents, including rollovers.

Also, cylinder 3 on the accident vehicle was fractured by overstress resulting from the initial impact of the front of the cylinder on the roadway or terrain during the rollover of the vehicle, and the cylinder was ejected from the semitrailer. The cylinder was particularly vulnerable to absorbing the initial impact with the roadway or ground because its body extended beyond the envelope of the bulkheads. The hazardous materials regulations are silent regarding the protection and shielding of horizontally mounted cylinders on semitrailers from initial impact during rollover accidents. According to the manufacturers of semitrailers with horizontally mounted cylinders, the accident semitrailer was typical of other semitrailers in service. Consequently, the Safety Board concludes that because horizontally mounted cylinders on semitrailers typically extend beyond the envelope of the bulkheads, the cylinders are exposed and vulnerable to initial impact with the roadway or terrain during rollover accidents and are at increased risk of damage, failure, and ejection. Therefore, the Safety Board believes that RSPA should require that cylinders that transport hazardous materials and are horizontally mounted on a semitrailer be protected from impact with the roadway or terrain to reduce the likelihood of their being fractured and ejected during a rollover accident.

Emergency Response

The Washington County 911 Center was timely in its notifications, and the emergency responders responded promptly. Even before the Washington County Emergency Management director had arrived on scene, he wisely requested help in identifying the hazards associated with compressed hydrogen.

The director stated that although the accident scene was stable when he arrived, the initial staging of personnel and emergency equipment was too close to the accident vehicle and had limited escape routes. The staging area was approximately 40 feet west of the overturned semitrailer, which was on fire. He considered that the risk of the situation destabilizing while firefighters broke down their equipment and moved to a safer location was greater than the risk posed by having the responders continue to spray large amounts of water on the cylinders to keep them cool. However, it would have been prudent for the director to relocate the equipment and the personnel not immediately essential to the cooling operation.

The firefighters did not move to a safer distance until 3:35 p.m., after a pumper engine from the Bartlesville fire department, unmanned ground monitors with the Phillips Petroleum DART, and the chief of the Tulsa hazardous materials team arrived. The chief and the director concluded that they could maintain the accident scene's stability while restaging equipment and personnel. They based their decision on the fact that additional equipment and the Tulsa hazardous materials team had arrived.

To address the staging problems at the accident scene, the director and Washington County have taken positive actions since the accident. One of the actions of Washington County has been a community-wide drill (fall 2001) that involved a simulated release of hazardous materials resulting in a simulated fire with multiple simulated injuries and fatalities. The drill included all fire departments within the county, as well as Phillips Petroleum. Also, each volunteer fire department within the county is performing drill scenarios.

In addition, Washington County established its own hazardous materials response team and is providing the team members with more advanced training in responding to hazardous materials accidents.

Emergency Response Guidance

The *North American Emergency Response Guidebook* is often the first reference used by the emergency responders who are first to arrive at an accident scene. The first responders to the Ramona accident used the book and referred to the guide for hydrogen. However, the guide did not provide complete information about the unique properties of hydrogen, specifically that hydrogen burns with an invisible, or almost invisible, flame. The guide also contained generic information about chemical properties, such as vapors sinking to the ground, which do not apply to hydrogen.

Incomplete or inaccurate information in the guidebook may lead first responders to take measures that endanger them. Consequently, the Safety Board concludes that although the incomplete or inaccurate information about hydrogen in the *North American Emergency Response Guidebook* was not a factor in this accident, there is the possibility that the lack of information could increase the risk to emergency response personnel.

Therefore, the Safety Board believes that RSPA should revise the information about hydrogen in the *North American Emergency Response Guidebook* so that it specifically identifies the unique chemical and flammability properties of hydrogen.

Conclusions

Findings

1. The following factors can be excluded as causal or contributory to the accident: highway conditions, weather, the mechanical condition of the vehicles, and impairment from alcohol and illegal drugs.

2. Because the pickup truck driver was unable to recall information about her activities, including her use of medications and sleep in the 96 hours before the accident, the Safety Board cannot determine with any confidence why her driving was erratic or why she lost control of her vehicle.

3. The truckdriver's operation of the tractor-semitrailer did not cause or contribute to the accident.

4. Because the valves, piping, and fittings at the rear of the semitrailer were not adequately protected and shielded from the impact caused by the rollover of the semitrailer, 8 of the 10 shutoff valves were sheared off, which resulted in the release and ignition of the hydrogen.

5. Cylinder 3 absorbed the initial impact with the roadway or terrain, resulting in the fracture of its front end and the ejection of the cylinder from the semitrailer.

6. The hazardous materials regulations do not provide sufficient and clear requirements for protecting cylinders and valves, piping, and fittings of cylinders that are horizontally mounted on semitrailers.

7. Because horizontally mounted cylinders on semitrailers typically extend beyond the envelope of the bulkheads, the cylinders are exposed and vulnerable to initial impact with the roadway or terrain during rollover accidents and are at increased risk of damage, failure, and ejection.

8. Although the incomplete or inaccurate information about hydrogen in the *North American Emergency Response Guidebook* was not a factor in this accident, there is the possibility that the lack of information could increase the risk to emergency response personnel.

Probable Cause

The National Transportation Safety Board determines that the probable cause of the May 1, 2001, collision and subsequent fire involving a tractor-semitrailer and a pickup truck in Ramona, Oklahoma, was the failure, for unknown reasons, of the pickup driver to control her vehicle. Contributing to the severity of the accident were the inadequate protection and shielding of the cylinders, valves, piping, and fittings and the inadequate securement of cylinders on the semitrailer.

Recommendations

As a result of its investigation, the National Transportation Safety Board makes the following safety recommendations:

To the Research and Special Programs Administration:

Modify 49 *Code of Federal Regulations* 173.301 to clearly require that valves, piping, and fittings for cylinders that are horizontally mounted and used to transport hazardous materials are protected from multidirectional forces that are likely to occur during accidents, including rollovers. (H-02-23)

Require that cylinders that transport hazardous materials and are horizontally mounted on a semitrailer be protected from impact with the roadway or terrain to reduce the likelihood of their being fractured and ejected during a rollover accident. (H-02-24)

Revise the information about hydrogen in the *North American Emergency Response Guidebook* so that it specifically identifies the unique chemical and flammability properties of hydrogen. (H-02-25)

BY THE NATIONAL TRANSPORTATION SAFETY BOARD

MARION C. BLAKEY
Chairman

CAROL J. CARMODY
Vice Chairman

JOHN A. HAMMERSCHMIDT
Member

JOHN J. GOGLIA
Member

GEORGE W. BLACK, JR.
Member

Adopted: September 17, 2002

this page intentionally left blank

Appendix A

Investigation

The National Response Center notified the Safety Board of the accident by facsimile in the early morning of May 2, 2001. The Safety Board dispatched a nine-person team (from Washington, D.C., and field offices) consisting of an investigator-in-charge, three highway investigators, two hazardous materials investigators, two human performance investigators, and one survival factors investigator. No Board Member participated in the on-scene investigation.

No hearings were held on the accident, and no depositions were taken.

Parties to the investigation were the Federal Motor Carrier Safety Administration, the Oklahoma Highway Patrol, the Tulsa Fire Department, the Washington County Emergency Management Agency, Airgas Mid-South, Inc., the Oklahoma Department of Transportation, and CP Industries, Inc. (The Safety Board also provided information about the investigation on an ongoing basis to the Research and Special Programs Administration.)

Appendix B

Emergency Response Timeline

TIME	ACTIONS
May 1	
2:15 p.m.	Accident occurs; 911 center is notified, Ramona volunteer fire department members are dispatched.
2:16 p.m.	Washington County Emergency Management director begins traveling to the accident scene.
Between 2:16 and 2:21 p.m.	On-scene witnesses extricate pickup driver from vehicle. Airgas Mid-South employee passing the accident notifies the company.
2:21 p.m.	Washington County Emergency Management director asks for assistance from the Phillips Petroleum dependable accident response team (DART).
2:22 p.m.	Chief of the Ramona volunteer fire department arrives on scene and assumes incident command. Washington County Emergency Management director notifies Ramona fire department responders that the cylinders might be pressurized and asks the fire departments from surrounding areas for mutual aid.
After 2:22 p.m.	Firefighters begin to spray water on the burning vegetation and the rear of the semitrailer and attempt to extricate the truckdriver from the tractor.
2:25 p.m.	Washington County Emergency Operations Center and Washington County Emergency Operations Plan are activated.
2:30 p.m.	Washington County Emergency Management director arrives and assumes incident command from the Ramona fire department chief. Oklahoma Highway Patrol officer trained in hazardous materials arrives on scene.
2:35 p.m.	Washington County Emergency Management director tells the firefighters to spray water on the cylinders from a distance. Shift superintendent of Phillips Petroleum DART tells the firefighters to stay away from the ends of the cylinders.
2:40 p.m.	Firefighters stop trying to extricate the truckdriver because of lack of water to continue cooling the cylinders.
2:55 p.m.	More water to cool the cylinders arrives; firefighters deploy more handlines to maintain cooling streams on the cylinders.

3:00 p.m.	Airgas president, vice president, and safety director begin traveling to accident scene.
3:11 p.m.	Oklahoma Highway Patrol officer trained in hazardous materials requests the assistance of the Tulsa fire department's hazardous materials response team.
3:15 p.m.	Firefighters extricate the truckdriver.
3:30 p.m.	Tulsa fire department's hazardous materials team begins to arrive.
3:35 p.m.	Chief of the Tulsa hazardous materials team assumes incident command and directs the firefighters to move away from the fire. He moves the fire suppression personnel across the highway and sets up system of water shuttles and unmanned monitors. Tulsa hazardous materials team surveys the condition of the cylinders' valves and monitors the cylinders' temperature.
4:00 p.m.	Airgas safety director offers the assistance of the Airgas AERO team to the chief of the Tulsa hazardous materials team.
5:15 p.m.	AERO team arrives to assist the firefighters with cooling and monitoring operations.
6:00 p.m.	Chief of the Tulsa hazardous materials team and the Washington County Emergency Management director consider the accident scene stabilized and discuss releasing the Tulsa fire department units.
6:30-6:40 p.m.	Tulsa fire department responders demobilize and leave the scene.
6:40 p.m.	Washington County Emergency Management director resumes incident command and directs Airgas to monitor cylinder temperature and the firefighters to continue cooling operations.
May 2	
12:20 a.m.	Airgas personnel advise that all cylinders are vented and fire is out.
12:55 a.m.	Washington County Emergency Management director leaves the scene and the Oklahoma Highway Patrol assumes incident command.
1:03 a.m.	Last fire and rescue units leave the scene.
6:00 a.m.	Highway reopens.

Appendix C

Safety Recommendations Issued as a Result of Highway Accident Report NTSB/HAR-01/01

To the Federal Motor Carrier Safety Administration:

H-01-17 through -25

Develop a comprehensive medical oversight program for interstate commercial drivers that contains the following program elements:

H-01-17

Individuals performing medical examinations for drivers are qualified to do so and are educated about occupational issues for drivers.

H-01-18

A tracking mechanism is established that ensures that every prior application by an individual for medical certification is recorded and reviewed.

H-01-19

Medical certification regulations are updated periodically to permit trained examiners to clearly determine whether drivers with common medical conditions should be issued a medical certificate.

H-01-20

Individuals performing examinations have specific guidance and a readily identifiable source of information for questions on such examinations.

H-01-21

The review process prevents, or identifies and corrects, the inappropriate issuance of medical certification.

H-01-22

Enforcement authorities can identify invalid medical certification during safety inspections and routine stops.

H-01-23

Enforcement authorities can prevent an uncertified driver from driving until an appropriate medical examination takes place.

H-01-24

Mechanisms for reporting medical conditions to the medical certification and reviewing authority and for evaluating these conditions between medical certification exams are in place; individuals, health care providers, and employers are aware of these mechanisms.

H-01-25

Develop a system that records all positive drug and alcohol test results and refusal determinations that are conducted under the U.S. Department of Transportation testing requirements, require prospective employers to query the system before making a hiring decision, and require certifying authorities to query the system before making a certification decision.

To the American Association of Motor Vehicle Administrators:

H-01-26

Urge your member States to develop a comprehensive medical oversight program for intrastate commercial drivers that contains the following program elements:

- Individuals performing medical examinations for drivers are qualified to do so and are educated about occupational issues for drivers.

- A tracking mechanism is established that ensures that every prior application by an individual for medical certification is recorded and reviewed.

- Medical certification regulations are updated periodically to permit trained examiners to clearly determine whether drivers with common medical conditions should be issued a medical certificate.

- Individuals performing examinations have specific guidance and a readily identifiable source of information for questions on such examinations.

- The review process prevents, or identifies and corrects, the inappropriate issuance of medical certification.

- Enforcement authorities can identify invalid medical certification during safety inspections and routine stops.

- Enforcement authorities can prevent an uncertified driver from driving until an appropriate medical examination takes place.

- Mechanisms for reporting medical conditions to the medical certification and reviewing authority and for evaluating these conditions between medical certification exams are in place; individuals, health care providers, and employers are aware of these mechanisms.

this page intentionally left blank